AF688178

Table of Contents

1 Introduction to Genetics

The very foundation of life, heredity, rests upon the elegant double helix of deoxyribonucleic acid, or DNA. This molecule, a master blueprint, directs the construction and function of every living organism. Imagine it as a complex instruction manual, meticulously detailing the creation of proteins, the workhorses of the cell. These proteins determine everything from our eye color to our susceptibility to certain diseases. Understanding DNA's structure—its sugar-phosphate backbone, its nitrogenous bases adenine, guanine, cytosine, and thymine, paired in their specific complementary fashion—is paramount to grasping the intricacies of genetics. The sequence of these bases forms the genetic code, a language that dictates the synthesis of proteins. This code, replicated faithfully during cell division, ensures the transmission of hereditary information across generations. Mistakes in this replication, however, can lead to mutations, the raw material of evolution and also the source of many genetic disorders.

Genes, the fundamental units of heredity, are specific segments of DNA that carry the instructions for building a particular protein or performing a specific function. Each gene occupies a precise location on a chromosome, a long, thread-like structure containing numerous genes. Humans, for example, possess 23 pairs of chromosomes, one set inherited from each parent. These chromosomes carry the complete set of an individual's genetic information, their genome. The combination of genes inherited from both parents determines an individual's genotype, while the observable characteristics—the physical traits, or phenotype—result from the interaction of the genotype with environmental factors. Environmental influences can modify gene expression, meaning that the same genotype can lead to different phenotypes under different conditions. This interaction is a key factor in the complexity of genetic inheritance.

Mendelian genetics, named after Gregor Mendel's groundbreaking work with pea plants, provides a foundational understanding of inheritance patterns. Mendel's experiments revealed the principles of segregation and independent assortment. Segregation states that during gamete formation (the creation of sperm and egg cells), each gene pair separates, resulting in each gamete receiving only one allele (a variant form of a gene). Independent assortment describes how different gene pairs assort independently of one another during gamete formation, contributing to the genetic diversity in offspring. Understanding these principles allows us to predict the probability of certain traits appearing in subsequent generations using Punnett squares and other genetic tools. Simple Mendelian inheritance patterns are crucial for understanding basic concepts, but many traits exhibit more complex inheritance patterns due to the involvement of multiple genes and environmental interactions.

Beyond simple Mendelian inheritance, the interplay between multiple genes (polygenic inheritance) and environmental factors (multifactorial inheritance) often complicates the relationship between

genotype and phenotype. For example, height and skin color are polygenic traits influenced by numerous genes, with environmental factors further shaping their expression. Understanding these more intricate inheritance patterns requires sophisticated statistical methods and advanced genetic analysis techniques. Furthermore, the concept of epistasis, where one gene affects the expression of another, adds another layer of complexity to genetic inheritance patterns. These interactions highlight the intricate network of genetic and environmental influences that shape an organism's characteristics. This understanding is essential for interpreting complex genetic data and deciphering the mechanisms underlying various traits and diseases. The field of genetics continues to evolve, uncovering increasingly nuanced interactions between genes and the environment, shaping our understanding of the complex processes of inheritance.
The study of human genetics has blossomed in recent decades, propelled by advances in molecular biology and genomics. These technological breakthroughs have enabled scientists to map the entire human genome, identifying the location and sequence of virtually every gene. This unprecedented level of detail has revolutionized our ability to diagnose, treat, and even prevent genetic disorders. Furthermore, this knowledge is driving the development of personalized medicine, tailoring treatments to an individual's unique genetic makeup. The ability to analyze an individual's genome provides a wealth of information for identifying disease risks, predicting drug response, and designing more effective therapies. It's a transformative shift in healthcare, moving away from a one-size-fits-all approach to a personalized approach focused on optimizing treatment based on an individual's genetic characteristics. The future of healthcare is

undeniably intertwined with the ongoing advancements in genomics and genetics.

1.1 DNA Structure and Function

The DNA molecule, the very blueprint of life, is a stunningly elegant structure. Imagine a twisted ladder, a double helix, where the sides are formed by alternating sugar and phosphate molecules. These form the backbone, the strong, stable framework holding everything together. But the real magic lies in the rungs of this ladder, the pairs of nitrogenous bases – adenine (A) always pairing with thymine (T), and guanine (G) with cytosine (C). This precise pairing is crucial; it dictates the genetic code, the sequence of instructions for building and maintaining an organism. The sequence of these bases, this specific order of A, T, G, and C, determines everything from eye color to susceptibility to certain diseases. Understanding this fundamental structure is the cornerstone of understanding genetics.

Its function is equally remarkable. DNA's primary role is to store and transmit genetic information. Think of it as a vast library, holding the complete set of instructions needed to build and run a living organism. This information isn't just passively stored; it's actively used. The process of gene expression involves transcription, where a section of DNA is copied into messenger RNA (mRNA), and translation, where the mRNA is used as a template to synthesize proteins. These proteins are the workhorses of the cell, carrying out a vast array of functions, from catalyzing biochemical reactions to providing structural support. The precise regulation of gene expression is critical, ensuring that the right proteins are produced at the right time and in the right amounts. This intricate dance of transcription and translation underpins the complexity of life itself. Mistakes in this process, mutations in the DNA sequence, can have profound consequences, ranging from minor variations in traits to serious genetic disorders.

Consider DNA replication, a critical process ensuring the accurate duplication of the genetic material during cell division. This process, mediated by specialized enzymes, meticulously unwinds the double helix, separates the two strands, and then synthesizes new complementary strands using each original strand as a template. This ensures that each new cell receives an identical copy of the genetic information. The accuracy of DNA replication is astonishing, with errors occurring only rarely. However, these rare errors, mutations, are the raw material of evolution, providing the variation upon which natural selection acts. The fidelity of DNA replication, therefore, represents a delicate balance between accuracy and the introduction of genetic variation.

Furthermore, the structure of DNA directly impacts its function. The double-helix configuration protects the genetic information from damage, shielding the vulnerable nitrogenous bases within its core. The specific pairing of bases ensures the accurate transmission of genetic information during replication and transcription. The antiparallel nature of the two strands, running in opposite directions, is crucial for the proper functioning of DNA replication enzymes. These structural features aren't arbitrary; they are intimately connected to the molecule's role in storing, transmitting, and expressing genetic information. They are beautifully designed, representing a perfect union of form and function.

Beyond its core functions of storage, transmission, and expression of genetic information, DNA plays a significant role in a range of cellular processes. For example, it is involved in DNA repair mechanisms that constantly monitor and correct errors that may arise from various sources, including exposure to radiation and other environmental factors. This continuous repair is essential for maintaining the integrity of the genome. Additionally, DNA is involved in the regulation of gene expression, allowing cells to respond to changing environmental conditions and internal signals. This dynamic regulation is crucial for development, differentiation, and the overall functioning of an organism. The study of DNA, therefore, is not just a study of a single molecule, but a study of the fundamental processes underlying life itself. It underpins all aspects of biology, from the smallest bacterium to the most complex organism.

This understanding of DNA structure and function is not just an academic pursuit; it's fundamental to numerous scientific and technological advancements. From genetic engineering techniques such as CRISPR-Cas9 to personalized medicine approaches that tailor treatments to an individual's genetic makeup, DNA's secrets are being unlocked to improve human health and well-being. The future holds even more promise as our understanding of DNA continues to deepen, opening up exciting possibilities for tackling complex diseases and improving human life in profound ways. However, with this power comes responsibility; navigating the ethical implications of these advancements is paramount to ensure that these powerful technologies are used for the betterment of humanity. The journey into the world of DNA is a journey into the very essence of life itself.

1.2 Genes and Inheritance

Genes, the fundamental units of heredity, are segments of DNA that code for specific traits. They are passed from parents to offspring through the process of sexual reproduction, a complex dance of chromosomes shuffling and pairing. This inheritance isn't always straightforward; some traits are dominant, masking recessive ones, while others exhibit more intricate patterns of inheritance, influenced by multiple genes or environmental factors. Understanding these patterns is key to comprehending the diversity of life and the potential for genetic manipulation.

Consider the simple case of eye color. It's often simplified as a single gene trait, but reality is far more nuanced. Multiple genes influence eye color, interacting in complex ways to produce a range of shades from deep brown to bright blue. The inheritance of such a seemingly simple trait highlights the intricate network of genetic interactions underlying even seemingly basic phenotypic variations. This complexity underscores the challenge, and yet the immense potential, of manipulating genes to affect specific traits.

The concept of alleles, different versions of the same gene, adds another layer to this complexity. You inherit two alleles for each gene, one from each parent. Your genotype, the combination of alleles you possess, determines your phenotype, the observable characteristic. A homozygous genotype has two identical alleles (e.g., two alleles for brown eyes), while a heterozygous genotype carries two different alleles (e.g., one allele for brown eyes and one for blue eyes). Dominant alleles express themselves even in the presence of a recessive allele, while recessive alleles require two copies for their expression. This fundamental principle is the foundation for understanding many in-

heritance patterns.
Beyond simple Mendelian inheritance, many traits are polygenic—influenced by multiple genes. Height, for instance, is not determined by a single gene but by the interplay of numerous genes, each contributing a small effect. Furthermore, environmental factors like nutrition and exercise also play significant roles in shaping the final phenotype. Recognizing the contribution of both genes and environment is crucial in understanding the complete picture of trait inheritance and variability. This understanding becomes even more critical when considering the ethical implications of genetic engineering and personalized medicine.
Imagine the implications of these principles in predicting the probability of inheriting a particular genetic disorder. Using Punnett squares, a simple tool in genetics, we can visualize the potential genotypes and phenotypes of offspring based on the parents' genotypes. This visualization becomes especially useful when dealing with recessive disorders, where carriers—individuals with one copy of a recessive allele—may not display the disorder themselves but can pass it on to their children. This knowledge is invaluable for genetic counseling and informed decision-making.
Understanding inheritance patterns extends beyond individual genes. Chromosomal abnormalities, involving changes in the number or structure of chromosomes, can have profound effects on an individual's phenotype. Down syndrome, for example, results from an extra copy of chromosome 21. Such chromosomal abnormalities are typically not inherited in a simple Mendelian fashion, but their impact highlights the importance of the overall chromosomal context in gene expression and phenotypic outcome. The study of these abnormal-

ities illustrates the far-reaching consequences of disruptions in the intricate balance of genetic material.

The field of genomics, the study of entire genomes, has revolutionized our understanding of inheritance. High-throughput sequencing technologies allow us to analyze an individual's entire DNA sequence quickly and cost-effectively, revealing a wealth of information about their genetic makeup, including variations that might predispose them to certain diseases or affect their response to specific treatments. This information is crucial for personalized medicine, which tailors medical care to an individual's unique genetic profile. The vast amount of genomic data, however, also presents substantial challenges in terms of data analysis, interpretation, and ethical considerations. Further developments in computational biology and bioinformatics are critical to tackling these challenges effectively.

In summary, inheritance is a complex and fascinating interplay of genes, alleles, chromosomes, and environmental factors. Understanding the basic principles of Mendelian inheritance, polygenic traits, and chromosomal abnormalities is essential for appreciating the intricacies of genetic diversity and the profound implications of genetic engineering technologies. The power to manipulate genes for therapeutic purposes is now within our reach, but this power carries significant responsibility, underscoring the importance of careful consideration of the ethical, societal, and environmental implications of such advancements.

2 Genetic Engineering Tools

Recombinant DNA technology forms the bedrock of many genetic engineering feats. Imagine splicing a gene from a firefly into a tobacco plant, resulting in a glowing plant. This seemingly simple act, achievable through techniques like restriction enzyme digestion and ligation, showcases the power of manipulating DNA. Scientists utilize specific enzymes to cut DNA at precise locations, creating sticky ends that readily combine with complementary sequences from other DNA fragments. These fragments, often carrying a gene of interest, are then joined using ligase enzymes, essentially sewing the genetic material together. The newly constructed DNA molecule is subsequently introduced into a host organism, like bacteria, where it replicates and expresses the introduced gene, thereby producing the desired protein or trait. This technique is not limited to plants; it's pivotal in producing therapeutic proteins like insulin for diabetics, creating disease-resistant crops, and generating genetically modified organisms for various research applications. Mastering this technique requires a thorough understanding of molecular biology principles, meticulous laboratory practices, and proficiency in various molecular biology tools, including PCR and gel electrophoresis for DNA manipulation and analysis.

Gene therapy, a field brimming with promise, focuses on correcting defective genes to treat inherited diseases. Different approaches exist, each tailored to the specific genetic defect. Viral vectors, modified viruses incapable of causing illness, are commonly used to deliver corrective genes directly into the patient's cells. These viruses, with their inherent ability to infect and integrate their genetic material into cells, serve as efficient delivery vehicles. However, challenges remain, particularly in ensuring long-term efficacy and minimizing potential side effects. Non-viral methods, such as lipid nanoparticles or direct DNA injection, present alternatives, albeit with limitations in their delivery efficiency. Careful consideration of the target cells, the type of genetic defect, and the potential risks is paramount. The efficacy of gene therapy is also influenced by factors like immune response and the integration site of the introduced gene. Ongoing research continuously improves gene therapy techniques, expanding its potential to treat a wider range of genetic disorders.

CRISPR-Cas9, often hailed as a revolutionary gene editing tool, has undeniably transformed the field of genetic engineering. This system, adapted from bacteria's natural defense mechanism against viruses, offers unparalleled precision and simplicity. The CRISPR-Cas9 system consists of two key components: a guide RNA molecule that directs the Cas9 enzyme to a specific target DNA sequence and the Cas9 enzyme, which acts as molecular scissors, cutting the DNA at the designated location. Once cut, the cell's natural repair mechanisms kick in, enabling researchers to either disrupt the targeted gene or precisely insert a new gene sequence. This technology has profound implications for treating genetic diseases, developing disease-resistant crops, and conducting basic research. Despite its incred-

ible potential, ethical concerns surrounding its use, particularly in germline editing, necessitate careful consideration and responsible implementation. Furthermore, the potential for off-target effects requires constant refinement and improvement of the technology.
Developing proficiency in these genetic engineering tools requires a multi-faceted approach. A strong foundation in molecular biology, genetics, and biochemistry is essential. Practical experience in the laboratory is crucial, allowing students to become proficient in the various techniques involved. This involves mastering techniques such as DNA extraction, PCR amplification, gel electrophoresis, cloning, and gene expression analysis. Beyond the laboratory, computational skills are becoming increasingly important. Bioinformatics tools allow researchers to analyze large datasets, design experiments, and interpret results. Finally, ethical considerations should be an integral part of any genetic engineering training, encouraging responsible application of these powerful technologies. The future of genetic engineering hinges on the ability of scientists to wield these tools responsibly while advancing scientific knowledge and benefiting society.

2.1 CRISPR-Cas9 Technology

CRISPR-Cas9, a revolutionary gene-editing technology, has rapidly transformed biological research and holds immense promise for treating genetic diseases. Its simplicity and precision are unparalleled, offering a powerful tool to modify DNA sequences with remarkable accuracy. Unlike previous gene-editing methods, CRISPR-Cas9 is relatively inexpensive, readily adaptable to various applications, and significantly easier to use, making it accessible to a broader range of researchers. This accessibility has fueled a surge in scientific breakthroughs, accelerating our understanding of gene function and paving the way for novel therapeutic interventions.

The system's core components are Cas9, a protein enzyme that acts as molecular scissors, and a guide RNA (gRNA) molecule. The gRNA acts as a GPS, precisely targeting a specific DNA sequence within the genome. Upon binding to the target DNA sequence, Cas9 creates a double-stranded break, initiating the cell's natural DNA repair mechanisms. These repair pathways can be manipulated to introduce specific genetic modifications, such as inserting, deleting, or replacing DNA segments. This ability to make precise changes within the genome opens up previously unimaginable possibilities.

One of the most exciting applications of CRISPR-Cas9 lies in its potential to cure genetic diseases. Many inherited disorders are caused by single gene mutations. CRISPR-Cas9 offers a direct approach to correct these mutations, potentially eliminating the disease at its source. For example, research is ongoing into using CRISPR-Cas9 to treat sickle cell anemia, cystic fibrosis, and Huntington's disease, among many others. While still in early stages for many applications, the successes achieved in preclinical models are incredibly encouraging, promising innovative therapeutic strategies. The technology's adaptability allows researchers to tackle a wide range of genetic disorders, offering a personalized approach to treatment.

Beyond disease treatment, CRISPR-Cas9 has found widespread applications in basic research. Scientists utilize it to study gene function by precisely disrupting or modifying specific genes and observing the resulting effects. This approach allows for a deeper understanding of complex biological processes and disease mechanisms. Furthermore,

CRISPR-Cas9 has found use in developing disease models, allowing scientists to study human diseases more effectively and ethically in animal models. These models help researchers test potential new therapies before human trials, increasing the efficiency and reducing the risks associated with drug development. This enhances the precision of experiments and the reliability of research outcomes.

However, the widespread adoption of CRISPR-Cas9 also necessitates careful consideration of ethical implications. Concerns exist regarding off-target effects—unintended modifications at sites other than the intended target. While significant advances have been made in improving the specificity of CRISPR-Cas9, minimizing these off-target effects remains a crucial area of research. The potential for misuse, such as creating genetically modified humans or developing bioweapons, is also a serious concern that necessitates careful ethical guidelines and regulations. Moreover, long-term consequences of gene editing remain largely unknown, warranting cautious and responsible application of the technology. International collaborations and stringent regulatory frameworks are essential to mitigate these risks.

Despite these challenges, the potential benefits of CRISPR-Cas9 are transformative. The technology's precision, versatility, and relative ease of use are revolutionizing fields such as medicine, agriculture, and biotechnology. As research continues to refine CRISPR-Cas9 and address the ethical considerations, its applications will only expand, offering groundbreaking solutions to a wide range of problems. The ongoing development of safer and more efficient gene-editing techniques and a thorough understanding of the ethical dimensions will ensure that the incredible potential of CRISPR-Cas9 is responsibly harnessed for the benefit of humanity, improving global health and addressing challenges confronting the world. Further research focusing on mitigating potential risks and fostering a comprehensive understanding of its long-term impact will be critical. Continuous monitoring and refinement are vital to ensuring the ethical use of such a powerful tool.

2.2 Gene Therapy Approaches

Gene therapy, a revolutionary field, aims to treat or prevent disease by correcting faulty genes. Unlike other genetic engineering techniques, its primary focus is therapeutic, directly addressing genetic defects responsible for various illnesses. This makes it a powerful tool with immense potential for improving human health. Consider cystic fibrosis, a debilitating lung disease caused by a mutation in a single gene. Gene therapy offers the possibility of correcting this mutation, alleviating the symptoms and improving the quality of life for affected individuals. Its applications extend far beyond single-gene disorders; researchers are actively exploring gene therapy's potential in treating complex diseases with multiple genetic components, including certain types of cancer and heart disease. The ongoing research is pushing the boundaries of what was once considered impossible, giving hope to millions affected by currently incurable conditions.

Several approaches exist within the realm of gene therapy, each employing unique strategies to deliver therapeutic genes into cells. One common method involves viral vectors, modified viruses that are harmless yet capable of carrying and delivering genes into target cells. These vectors, such as adeno-associated viruses (AAV) or retroviruses, have been extensively studied and modified for optimal safety and efficacy. They're chosen for their ability to efficiently infect specific cell types, ensuring the therapeutic gene reaches its intended destination. However, the use of viral vectors does present challenges; potential immunogenicity, the body's immune response to the virus, and the limited cargo capacity of some vectors are key limitations being addressed through ongoing research. The search for better, safer and more efficient vectors remains a central focus within the field.

Alternatively, non-viral methods offer a different approach. These strategies, typically employing techniques like lipid nanoparticles or physical methods like electroporation, don't involve using viruses to deliver the therapeutic gene. While offering potential advantages regarding safety and immunogenicity, non-viral methods often struggle with less efficient gene delivery compared to their viral counterparts. The efficiency of gene transfer is often lower, requiring optimization and further research to enhance delivery rates. However, continuous improvements in technology and techniques provide optimism for their future use, especially in applications where viral vectors carry higher risks. This continuous development in delivery mechanisms is critical to the ongoing success and expansion of gene therapy.

Another crucial aspect involves choosing the appropriate therapeutic gene. Identifying and selecting the right gene to correct or compensate for the underlying genetic defect is paramount. This often requires an in-depth understanding of the specific disease mechanism and the role of the faulty gene. Scientists may modify the therapeutic gene to enhance its expression or stability within the target cells, optimizing its effectiveness. This careful selection and engineering of the therapeutic gene contribute significantly to the overall success of the gene therapy treatment. Identifying the correct therapeutic gene often involves extensive research, studying the genetic underpinnings of the disease, and determining the most suitable replacement or corrective gene.

The successful delivery and expression of the therapeutic gene aren't enough; ensuring long-term therapeutic effects remains a significant hurdle. Many gene therapy strategies focus on achieving sustained gene expression to maintain the therapeutic effect over time. This is a challenging aspect, as gene expression can be affected by various factors, including epigenetic modifications and immune responses. Researchers are exploring strategies such as using modified promoters, sequences that regulate gene expression, or alternative delivery systems to improve the longevity of therapeutic gene expression. Ultimately, achieving long-term therapeutic benefit is key to providing a lasting improvement in patients' lives and to fully realize the potential of gene therapy. This requires careful consideration of not only the initial delivery and expression, but also the long-term stability and function of the introduced gene.

Ethical considerations are paramount in the development and application of gene therapy. The potential to modify the human genome raises questions about the safety, efficacy, and potential unintended consequences of this powerful technology. Thorough preclinical testing and rigorous clinical trials are necessary to ensure the safety and efficacy of gene therapies before widespread use. Addressing concerns regarding equitable access and preventing potential misuse of the technology are also crucial aspects requiring ongoing discussion and ethical frameworks. These discussions involving scientists, ethicists, and the public are vital to establishing responsible guidelines

and practices for the application of gene therapy.
Despite challenges, gene therapy's progress is remarkable. Clinical trials have demonstrated its effectiveness in treating several inherited disorders, offering hope for patients with previously untreatable conditions. Further research, focused on enhancing gene delivery systems, improving therapeutic gene design, and addressing ethical considerations, will continue to expand the therapeutic scope of gene therapy. The future holds the potential for gene therapy to become a standard treatment option for a wider range of diseases, transforming the landscape of medicine and improving the quality of life for countless individuals. The remarkable progress and ongoing research create a sense of hope and excitement, marking gene therapy as one of the most promising advancements in modern medicine.

2.3 Recombinant DNA Technology

Recombinant DNA technology is a cornerstone of modern biotechnology, enabling scientists to manipulate DNA sequences and create novel combinations of genetic material. This process involves isolating a specific gene from one organism and inserting it into the DNA of another organism, a technique with far-reaching applications in medicine, agriculture, and industry. Imagine the possibilities: producing insulin for diabetics using genetically modified bacteria, creating disease-resistant crops, or developing new enzymes for industrial processes. These are just a few examples of the transformative power of recombinant DNA technology. The core principle relies on the universality of the genetic code, where the same codons specify the same amino acids across diverse species. This fundamental similarity allows genes from vastly different organisms to be expressed

successfully in a new host.
The creation of recombinant DNA involves several crucial steps, starting with gene isolation. Researchers use restriction enzymes, essentially molecular scissors, to cut DNA at specific sequences, isolating the gene of interest. These precise cuts create sticky ends, short stretches of single-stranded DNA, which facilitate the insertion of the target gene into a vector. Vectors, commonly bacterial plasmids or viruses, act as delivery vehicles, carrying the gene into the host organism. Ligases, enzymes that join DNA fragments, are then employed to covalently link the gene to the vector's DNA, completing the construction of the recombinant DNA molecule.
The next critical step involves introducing this recombinant DNA molecule into the host organism. This process, termed transformation or transduction depending on the vector used, relies on various techniques, including heat shock, electroporation, or viral infection. Once inside the host, the recombinant DNA replicates along with the host's genome, ensuring that subsequent generations of the host organism also carry the inserted gene. This ensures stable expression of the desired protein or trait. The successful transformation is verified through several techniques, including PCR, which amplifies the specific DNA sequence, and gene expression analysis, demonstrating the functional activity of the newly introduced gene.
The implications of recombinant DNA technology are profound and multifaceted. In medicine, it has revolutionized the production of therapeutic proteins, such as insulin, human growth hormone, and clotting factors. These proteins, previously obtained from scarce animal sources, are now manufactured in large quantities using genetically modified microorganisms, ensuring a consistent and safe sup-

ply for patients. Furthermore, recombinant DNA technology plays a vital role in gene therapy, offering the potential to correct genetic defects responsible for inherited diseases. Though still under development, the implications for treating cystic fibrosis, hemophilia, and other genetic disorders are transformative.

Beyond medicine, recombinant DNA technology has revolutionized agriculture. Genetically modified (GM) crops, engineered to resist pests, herbicides, or improve nutritional content, have increased crop yields and reduced pesticide use. However, the use of GM crops has sparked significant debate, raising concerns about environmental impacts, biodiversity, and potential health risks. These concerns necessitate ongoing research and careful regulation to ensure the responsible development and application of GM technologies. Likewise, the industrial applications of recombinant DNA technology are vast. The production of enzymes with improved stability, activity, or specificity has facilitated significant advancements in various sectors, including the food, textile, and pharmaceutical industries.

Mastering recombinant DNA technology requires a solid understanding of molecular biology, genetics, and microbiology. However, the practical applications of this technology are constantly expanding, fueled by ongoing advancements in gene editing tools like CRISPR-Cas9, high-throughput screening techniques, and synthetic biology approaches. As we continue to explore the potential of genetic engineering, recombinant DNA technology remains a cornerstone for innovation across multiple disciplines. It represents a powerful tool to address global challenges, such as food security, disease prevention, and environmental remediation. However, alongside its remarkable potential lies the responsibility to address ethical considerations and ensure that these powerful technologies are used responsibly and sustainably. The future of biotechnology rests, in part, on our ability to harness the power of recombinant DNA technology while mitigating potential risks.

3 Personalized Medicine

Understanding your own genetic makeup is the cornerstone of personalized medicine. It's not a one-size-fits-all approach; instead, it tailors treatments and preventative strategies to your unique genetic profile. This personalized approach considers your specific genetic variations that influence your susceptibility to diseases, your response to medications, and even your overall health trajectory. It's a paradigm shift from traditional medicine, where treatments are often generalized for large populations. The power lies in understanding how your genes uniquely interact with your environment to shape your individual health outcomes, far exceeding the efficacy of blanket treatments.

Pharmacogenomics, a critical component, studies how genes affect the way you metabolize medications. Different people process drugs differently, leading to vastly different therapeutic effects and side effects. Your genetic code helps determine which medication will be most effective and minimizes the risk of adverse reactions. For example, certain genetic variations can predict whether you'll experience a severe reaction to a particular drug, allowing doctors to choose an alternative treatment before any problems arise. This predictive capacity significantly enhances patient safety and treatment effectiveness; it's not about simply reacting to complications, but actively preventing them based on your genetic predisposition. This is not mere theory; it's actively saving lives and improving healthcare outcomes.

Cancer genomics utilizes genetic information to diagnose and treat cancers with unprecedented precision. Analyzing a tumor's genetic makeup reveals specific mutations driving its growth, pinpointing the most effective targeted therapies. For example, certain cancers are driven by specific mutations that can be targeted by precisely designed medications, enhancing treatment response and reducing the need for harsh, general chemotherapy. Furthermore, understanding a tumor's genetic profile can predict its likely behavior and progression, informing treatment strategies and long-term prognosis. This approach reduces unnecessary treatments and tailors care to suit the unique genetic profile of the individual cancer. This precision has radically transformed cancer treatment from a broad-brush approach to a finely tuned, personalized strategy.

Beyond cancer and drug response, personalized medicine extends to numerous other areas. Genetic testing can predict an individual's susceptibility to various diseases like diabetes, heart disease, or Alzheimer's. This information empowers individuals to make informed lifestyle choices and implement preventative measures, significantly mitigating their risk. Early intervention, guided by genetic insights, becomes a cornerstone of proactive healthcare. It's not just about treatment; it's about empowering individuals to take control of their health through informed decision-making based on their unique genetic landscape. Knowing your genetic predispositions allows you to proactively adjust your lifestyle – diet, exercise, and environmen-

tal exposure – to minimize risks and maximize your healthspan.
The implications of personalized medicine are far-reaching. Imagine a future where healthcare is not a reactive response to illness, but a proactive endeavor to prevent disease altogether. A future where treatments are precisely tailored to individual genetic profiles, optimizing effectiveness and minimizing side effects. The cost-effectiveness is another significant advantage. By targeting treatments only to those who will actually benefit, healthcare resources are used more efficiently. This efficient resource allocation is vital in ensuring broader access to personalized medicine for a wide range of patients. It reduces wasted resources and ultimately makes healthcare more sustainable. The future of healthcare rests on the ability to understand the language of our genes and tailor our approaches accordingly. The potential for a healthier and longer life, through personalized medicine, is profound.

3.1 Pharmacogenomics

Pharmacogenomics delves into the fascinating interplay between an individual's genetic makeup and their response to medications. It's not just about whether a drug works; it's about how effectively it works, and crucially, whether it causes adverse reactions. Understanding this intricate relationship is paramount for optimizing treatment and minimizing harm. This field holds immense promise for personalized medicine, offering tailored therapies based on your unique genetic profile.
The cornerstone of pharmacogenomics lies in identifying specific genes that influence drug metabolism. For example, variations in genes encoding enzymes involved in drug breakdown can significantly

impact how quickly a medication is processed by your body. Some individuals might metabolize a drug too quickly, rendering it ineffective, while others might metabolize it too slowly, leading to drug accumulation and potentially toxic side effects. This knowledge allows healthcare professionals to adjust dosages accordingly, ensuring optimal therapeutic effects while mitigating risks.

Consider the case of warfarin, a widely prescribed anticoagulant. Variations in the CYP2C9 and VKORC1 genes profoundly influence how the body metabolizes and responds to warfarin. Patients with certain gene variants require lower doses to achieve the desired therapeutic effect, whereas others might need higher doses. Pharmacogenomic testing can identify these variations, enabling clinicians to prescribe the right dose from the start, minimizing the risk of bleeding or clotting complications. This tailored approach enhances treatment efficacy and safety.

Beyond dosage adjustments, pharmacogenomics also helps predict adverse drug reactions. Some individuals carry genetic variations that increase their susceptibility to specific side effects. For instance, certain genetic variants can predispose individuals to a life-threatening reaction to certain medications. Pharmacogenomic testing can identify these predispositions, allowing healthcare professionals to avoid prescribing potentially harmful drugs and select safer alternatives. This preventative approach is critical for patient safety.

However, implementing pharmacogenomics into routine clinical practice presents its own set of challenges. The cost of pharmacogenomic testing can be prohibitive for some patients, and the interpretation of complex genetic data requires specialized expertise. Furthermore, the sheer volume of genetic information can be overwhelming, creating the need for user-friendly tools and guidelines to facilitate the integration of pharmacogenomic data into clinical decision-making. The development of robust clinical guidelines and standardized reporting practices is crucial to maximize the benefit of pharmacogenomic testing.

Despite these hurdles, the potential benefits of pharmacogenomics are undeniable. Imagine a future where every patient receives a medication regimen tailored precisely to their genetic makeup. This personalized approach could revolutionize healthcare, improving treatment outcomes and reducing adverse events. It's no longer a futuristic dream; it's a reality progressively becoming more accessible. The ongoing research and technological advancements are steadily reducing the costs and complexity of pharmacogenomic testing, making this revolutionary approach more practical. The combination of genetic testing and sophisticated data analysis is creating a new era in personalized medicine.

Pharmacogenomics, in its essence, is a powerful tool for optimizing drug therapy. By integrating an individual's genetic information into treatment decisions, we can move beyond a one-size-fits-all approach to medicine. This precision medicine approach has the potential to significantly improve patient outcomes and transform how we approach healthcare, moving from reactive treatment to proactive prevention. The future holds even greater promise as our understanding of the complex interplay between genes and drug response continues to evolve. This evolution promises to lead to even more effective and safer treatments. The journey toward truly personalized medicine is underway, with pharmacogenomics leading the charge. This personalized approach promises to improve the lives of millions and represent a significant leap forward in medical treatment. The impact on healthcare will be substantial, significantly altering the way medicine is practiced and experienced.

3.2 Cancer Genomics

Cancer is a complex disease, driven by alterations in a cell's DNA. These changes, or mutations, can affect genes that regulate cell growth, division, and death. Understanding these genetic alterations is crucial for developing effective, personalized cancer treatments. This is the core of cancer genomics.

Traditional cancer treatments, such as chemotherapy and radiation, often target rapidly dividing cells, impacting both cancerous and healthy cells, leading to significant side effects. Cancer genomics offers a more precise approach. By analyzing a tumor's DNA, researchers can identify the specific genetic mutations driving its growth. This information allows for the selection of targeted therapies that specifically inhibit the activity of mutated genes or proteins, minimizing harm to normal cells.

Imagine a patient diagnosed with lung cancer. Standard treatments might involve harsh chemotherapy. However, through genomic sequencing of the tumor, we discover a specific mutation, EGFR, frequently found in lung cancer. This finding points to a targeted therapy, a drug designed to specifically block the activity of the mutated EGFR protein. This personalized approach increases the chance of success and reduces the side effects compared to a more generalized treatment.

The application of cancer genomics goes beyond treatment selection. It also plays a significant role in early detection and diagnosis. By analyzing blood samples, for example, researchers can identify circulating tumor DNA (ctDNA), fragments of DNA shed by cancer cells into the bloodstream. Detecting ctDNA can indicate the presence of cancer even before any visible symptoms appear, offering an opportunity for early intervention and improved outcomes. This is particularly significant for cancers that are difficult to diagnose at early stages.

Furthermore, cancer genomics is revolutionizing prognosis. By analyzing the genetic profile of a tumor, oncologists can predict how aggressive the cancer is likely to be and how it will respond to various therapies. This information allows for more informed treatment decisions, helping to personalize treatment plans to better suit each patient's unique circumstances.

The information gleaned from cancer genomics also fuels research into novel treatment strategies. By identifying the genetic drivers of drug resistance, researchers can develop new approaches to overcome these resistances. This is a dynamic field, constantly evolving with new discoveries leading to new therapeutic targets and improved treatment strategies. For instance, the identification of specific gene fusions, like the EML4-ALK fusion, has led to the development of ALK inhibitors, substantially improving outcomes for certain lung cancer patients.

Beyond individual patient treatment, cancer genomics is instrumental in understanding the complexities of cancer at a broader level. Large-scale genomic studies, analyzing the genetic profiles of thousands of tumors across various cancer types, are helping scientists understand the intricate genetic landscapes of these diseases. These studies uncover common mutations, identify new therapeutic targets, and reveal underlying mechanisms driving cancer development and progression. This is a crucial step towards developing more effective and universal preventative strategies.

The ongoing advancements in sequencing technologies and computational biology are rapidly accelerating progress in this field. Faster and cheaper sequencing methods allow for broader access to cancer genomic testing. The development of powerful computational tools enhances the analysis of complex genomic data, revealing patterns and relationships between mutations and clinical outcomes. This is leading to a much deeper understanding of the intricate network of genetic interactions driving cancer.

However, challenges remain. Interpreting the vast amount of genomic data generated can be complex, requiring specialized expertise and sophisticated computational methods. The cost of genomic testing can be prohibitive in certain healthcare settings, limiting access for many patients. Ensuring equitable access to cancer genomic testing and therapies is a crucial aspect of responsible implementation of this technology. This requires collaborative efforts among researchers, clinicians, policymakers, and healthcare systems worldwide. Only through such collaborative approaches can we fully realize the potential of cancer genomics to revolutionize the way we understand, diagnose, and treat this devastating disease. The future of cancer treatment undoubtedly relies on continuing advancements and widespread application of cancer genomics.

4 Genetic Testing and Screening

Prenatal genetic testing offers expectant parents invaluable information about their unborn child's genetic makeup. Techniques like amniocentesis and chorionic villus sampling (CVS) allow for the detection of chromosomal abnormalities such as Down syndrome, trisomy 18, and trisomy 13. These procedures, while offering crucial insights, do carry a small risk of miscarriage, a factor parents must carefully weigh against the potential benefits of early diagnosis. Understanding these risks and the implications of the results is paramount before proceeding. The emotional impact of receiving positive or negative results should also be considered, highlighting the importance of genetic counseling to navigate these complex decisions. Choosing the appropriate test, understanding the limitations of each technique, and being fully informed about the potential consequences are critical steps in responsible prenatal care. Early detection can allow for preparation and planning, offering parents the opportunity to make informed choices regarding their pregnancy.

Carrier screening, on the other hand, focuses on identifying individuals who carry recessive genes for specific inherited disorders. This is particularly relevant for couples planning a family, as two carriers of the same recessive gene have a chance of having a child affected by the disorder. Commonly screened conditions include cystic fibrosis, sickle cell anemia, and Tay-Sachs disease. This type of testing allows couples to assess their risk of having an affected child and to make informed reproductive decisions. Options may range from choosing to conceive naturally, understanding the risks, to utilizing assisted reproductive technologies like preimplantation genetic diagnosis (PGD) to select embryos free of the disease-causing genes. Genetic counselors play a critical role in explaining the test results and guiding couples through the often emotionally charged decision-making process. Knowledge is power, and carrier screening empowers couples to make conscious choices aligned with their family planning goals.

Predictive genetic testing aims to assess an individual's risk of developing certain diseases later in life. This type of testing looks for genetic variations associated with an increased likelihood of conditions like breast cancer, Huntington's disease, or Alzheimer's disease. It's crucial to remember that these tests don't provide definitive diagnoses; rather, they offer risk probabilities. A positive result doesn't guarantee that a person will develop the disease, while a negative result doesn't eliminate the possibility entirely. Lifestyle factors and environmental influences also play significant roles in disease development. Therefore, predictive testing should be viewed as one piece of the puzzle, providing information that can inform lifestyle choices and preventative measures. Understanding the limitations of the test and the potential psychological implications of receiving a positive result are essential considerations. The power of this knowledge needs to be coupled with proper guidance and support.

Newborn screening is a critical public health initiative that involves testing newborns for a panel of genetic disorders shortly after birth. These tests are often mandated by law and are designed to detect conditions that can be treated effectively with early intervention. Examples include phenylketonuria (PKU) and congenital hypothyroidism. Early diagnosis through newborn screening allows for timely treatment, preventing severe complications and improving the child's long-term health outcomes. The screening process is typically non-invasive, involving a simple blood test. The vast majority of newborns receive these tests without any complications. The benefits significantly outweigh the minimal risks, making it a cornerstone of preventative healthcare. This early detection ensures that children receive appropriate medical care from the outset, enhancing their chances for a healthy future. This critical element of healthcare has positively impacted countless lives.

Direct-to-consumer (DTC) genetic tests have gained significant popularity in recent years, offering individuals access to genetic information directly without the involvement of a healthcare professional. These tests can provide information about ancestry, carrier status for certain diseases, and even predispositions to various health conditions. However, the interpretation of results from DTC tests can be challenging and may require additional guidance from a genetic counselor to ensure accuracy and proper understanding. Moreover, the regulatory oversight of DTC testing varies across jurisdictions, raising concerns about the quality and accuracy of the tests and the potential for misinformation. Responsible use of DTC tests necessitates a critical evaluation of the test provider's reputation, a clear understanding of the limitations of the test, and a cautious approach to interpreting the results. Consumers must seek professional guidance if they have questions or concerns about the results. These readily accessible tests offer individuals a glimpse into their genetic landscape, but responsible use is paramount to prevent potential misinterpretations and unwarranted anxiety.

4.1 Prenatal Genetic Testing

Prenatal genetic testing offers expectant parents invaluable information about their developing child's genetic makeup. It's a powerful tool, but one that requires careful consideration of its implications. These tests aren't without limitations; they can provide incomplete information, and the results can be emotionally challenging to process. Understanding the different types of prenatal testing, their accuracy rates, and the potential consequences is essential for informed decision-making.

Several testing methods exist, each with its own strengths and weaknesses. Ultrasound, a relatively non-invasive procedure, can reveal structural abnormalities in the fetus. However, it's less precise in detecting genetic disorders. Chorionic villus sampling (CVS), a more invasive procedure involving a sample from the placenta, carries a slightly higher risk of miscarriage but can detect chromosomal abnormalities earlier in pregnancy than amniocentesis. Amniocentesis, which involves extracting amniotic fluid, offers a more comprehensive genetic analysis but is performed later in the pregnancy. Non-invasive prenatal testing (NIPT) is a blood test for the mother, analyzing fetal DNA circulating in her bloodstream. While generally safer, NIPT might miss some abnormalities or provide false positives, necessitating follow-up testing. The choice of method depends on factors such as gestational age, risk factors, and the parents' preferences. The healthcare provider plays a crucial role in guiding the selection process, providing a detailed explanation of each procedure's advantages, disadvantages, and risk factors.

The information obtained from prenatal genetic testing can profoundly impact a pregnancy. Positive results may lead to difficult decisions, including the possibility of termination. Genetic counselors provide vital support, helping couples understand the results, weigh their options, and navigate the complex emotional landscape these tests can present. Furthermore, these counselors facilitate communication between parents and medical professionals, fostering a collaborative approach to managing the pregnancy and planning for the future.

It's crucial to remember that a negative result doesn't guarantee a completely healthy baby; some genetic conditions might be missed by current testing technologies. Conversely, a positive result for a particular condition doesn't always translate to a severe or life-threatening illness; many genetic disorders have varying degrees of severity. The information obtained helps parents prepare for the challenges a child with genetic conditions may face, allowing for early intervention and support. This preparation is paramount in enabling better care and improving the quality of life for the child and the family.

Access to prenatal genetic testing varies across geographical locations and healthcare systems. Financial constraints can restrict access for some families, highlighting the need for equitable access to these crucial services. Additionally, cultural and religious beliefs influence attitudes toward prenatal testing and subsequent decisions. Open communication between healthcare providers, genetic counselors, and patients is vital to ensure that individuals make informed choices aligned with their values and beliefs. These discussions must navigate complex ethical considerations, guaranteeing the respect of individual autonomy and promoting responsible medical practices. Furthermore, the development and widespread accessibility of low-cost, accurate, and reliable genetic testing are crucial to enhance population-level health outcomes and reduce health disparities.

Beyond the immediate implications for a particular pregnancy, the information gleaned from prenatal genetic testing can have long-term consequences for the family. Parents might undergo further genetic testing to assess their own carrier status or the risk of future pregnancies being affected by similar conditions. Such information empowers families to make proactive healthcare decisions and provides a basis for appropriate medical management. Understanding familial genetic predispositions can greatly inform family planning, contributing to a broader understanding of the family's overall health profile. It is a matter of great significance that the ethical implications of these long-term consequences are discussed openly and frankly with the patients during the prenatal testing consultation.

Furthermore, the continuous advancement of genetic testing technologies necessitates ongoing education and training for healthcare professionals. As new tests emerge and diagnostic capabilities improve, healthcare providers must stay updated on the latest advancements to ensure the provision of the best possible care. Medical professionals should receive continuous training to keep their skill sets relevant, allowing them to efficiently utilize the ever-evolving technologies and to correctly interpret the constantly increasing amounts of data. It is vital for healthcare systems to prioritize continuous professional development in this area to ensure the delivery of high-quality care that incorporates all the most recent medical breakthroughs.

4.2 Carrier Screening

Carrier screening offers prospective parents invaluable information about their genetic makeup. Understanding this process is crucial for making informed reproductive choices. It's not about eliminating "defective" genes; rather, it empowers couples to understand their risks and plan accordingly. This knowledge can be incredibly empowering, allowing for proactive decision-making regarding family planning. The emotional impact of these tests is substantial, and understanding this aspect is as important as understanding the science.

Many inherited conditions are recessive, meaning a child needs two copies of a faulty gene – one from each parent – to develop the disease. Carrier screening identifies individuals who carry one copy of a recessive gene but do not show symptoms themselves. These individuals are "carriers," and while they're healthy, they can pass the gene onto their children. Knowing carrier status allows couples to assess their combined risk of having a child with a specific genetic disorder. This information is not deterministic; it provides probabilities, not certainties.

The range of conditions screened for varies. Some tests focus on a single gene, such as the cystic fibrosis gene. Others are broader panels encompassing dozens of genes associated with numerous inherited diseases. Choosing the right test depends on several factors, including family history, ethnicity, and individual preferences. Genetic counselors are invaluable resources in this process, helping individuals navigate the complexities of testing and interpreting results. They can explain the science clearly and address any concerns.

Preparing for carrier screening involves understanding what to expect. The process typically involves a simple blood or saliva sample. Results are usually available within a few weeks. Pre-test counseling helps individuals understand the implications of the test and what the results might mean. Post-test counseling is equally important, providing support and guidance regardless of the results. Whether the results reveal carrier status or not, the process can trigger complex emotions.

The information obtained through carrier screening is personal and confidential. It should only be shared with those individuals who need to know, typically the couple involved. Protecting the privacy of this genetic information is crucial, given the potential implications for insurance and employment. This protection is vital for preserving individual autonomy and preventing discrimination based on genetic predisposition. Understanding the legal and ethical considerations surrounding genetic testing is paramount.

Beyond the individual level, carrier screening has broader implications for public health. Population-level data gathered from carrier screening can help scientists better understand the prevalence of genetic disorders in different populations. This data is essential for public health initiatives, including improved disease surveillance and the development of targeted prevention strategies. The information contributes to the larger field of genetic epidemiology, improving our collective understanding of inherited conditions. Understanding this broad impact is essential.

The cost of carrier screening varies depending on the specific test and the number of genes analyzed. Insurance coverage also differs. Some insurance plans fully cover carrier screening, particularly if there is a family history of a specific condition. Others may require pre-authorization or only cover certain tests. Exploring insurance coverage options is crucial prior to testing, to avoid unexpected financial burdens. Understanding financial implications is crucial.

The decision of whether or not to undergo carrier screening is a deeply personal one. It's vital to thoroughly consider the potential benefits and limitations. It's important to weigh the possibility of learning about a carrier status with the emotional and potential reproductive implications against the benefits of obtaining this information. A fully informed decision is paramount. The ultimate goal is to use this information to support reproductive choices consistent with personal values and desires. This should be empowering and provide clarity. Ultimately, this choice is a deeply personal one, and there is no right or wrong answer.

4.3 Predictive Genetic Testing

Predictive genetic testing offers a glimpse into the future, allowing individuals to assess their risk of developing certain diseases before symptoms appear. This proactive approach empowers individuals to make informed decisions about their health, potentially preventing or delaying the onset of debilitating conditions. However, it's crucial to understand the complexities and limitations of this technology. The results aren't always definitive; a positive result doesn't guarantee disease development, and a negative result doesn't eliminate all risk. Furthermore, the emotional impact of receiving predictive genetic information can be significant, requiring careful counseling and support.

Many predictive tests focus on single-gene disorders, such as Huntington's disease or cystic fibrosis, where a specific genetic mutation directly causes the disease. For these conditions, testing provides a relatively clear picture of an individual's risk. However, the majority of common diseases, like heart disease, diabetes, and certain cancers, are polygenic, meaning they are influenced by multiple genes and environmental factors. Predictive testing for these complex diseases is more challenging, often providing only a probability of developing the condition, rather than a definitive yes or no. These probabilistic results demand careful interpretation and consideration of other risk factors.

The availability of predictive genetic testing is constantly expanding, encompassing a broader range of conditions. Technological advancements are making the process more efficient and cost-effective, leading to increased accessibility. However, this also raises concerns about the potential for misuse and misinterpretation of results. It's essential for individuals considering predictive testing to have a thorough understanding of the test's limitations and potential implications before proceeding. Genetic counselors play a vital role in this process, providing personalized guidance and support to ensure informed decision-making.

Beyond the individual level, predictive genetic testing has implications for families and populations. Identifying individuals at high risk for certain diseases can facilitate early interventions, lifestyle modifications, and preventative measures. This can significantly impact the overall burden of disease and improve public health outcomes.

However, access to such testing needs to be equitable and affordable to avoid creating further health disparities. Furthermore, the ethical implications of widespread predictive testing should be carefully considered, particularly concerning issues of privacy, discrimination, and the potential for undue stress or anxiety.

Consider, for example, the implications of a predictive test for Alzheimer's disease. Receiving a positive result might significantly impact an individual's life choices, from career planning to retirement decisions. The emotional and psychological burden can be substantial, highlighting the critical need for comprehensive support systems and resources for individuals undergoing predictive genetic testing. Conversely, a negative result may provide reassurance but doesn't guarantee a disease-free future. Thus, the interpretation and management of the results should be a collaborative process involving the individual, their healthcare provider, and potentially a genetic counselor.

Therefore, while predictive genetic testing offers a powerful tool for proactive healthcare management, its application requires careful consideration of its limitations, ethical implications, and psychological impact. The future of predictive genetic testing depends on continued research to improve test accuracy and accessibility while simultaneously addressing ethical concerns and providing comprehensive support for individuals undergoing testing. This includes ensuring equitable access to testing and counseling, developing robust frameworks for data privacy and security, and providing ongoing education and support to promote responsible use and interpretation of predictive genetic information. The potential benefits are vast, but responsible implementation is crucial for maximizing these benefits while mitigating potential risks.

4.4 Newborn Screening

Newborn screening is a crucial public health initiative, offering a vital window into a baby's genetic makeup shortly after birth. It involves analyzing a small blood sample, typically taken from the infant's heel, to detect a range of treatable conditions. These conditions, often undetectable without screening, can cause severe developmental delays or even death if left untreated. Early identification through this program allows for immediate interventions, significantly improving the child's outcome. The process is relatively painless and non-invasive for the newborn, making it a widely accepted practice.

The specific conditions screened vary by state and country, reflecting evolving scientific understanding and technological advancements. Many programs include tests for phenylketonuria (PKU), a metabolic disorder affecting the breakdown of phenylalanine; congenital hypothyroidism, a condition affecting thyroid hormone production; and cystic fibrosis, a genetic disorder affecting the lungs and digestive system. The inclusion of specific conditions is often based on a combination of factors, such as the prevalence of the disorder, the severity of its effects, and the availability of effective treatments. These assessments are continually reevaluated to ensure the most effective and efficient use of resources.

Myriad factors influence the success and effectiveness of newborn screening programs. Accuracy of testing is paramount, requiring meticulous laboratory procedures and quality control measures. Rapid turnaround time for results is vital; delays can hinder timely intervention. Equally important is the accessibility of screening to all newborns, regardless of socioeconomic status or geographic location. This requires robust infrastructure, including well-trained personnel, adequate laboratory facilities, and effective communication channels to disseminate results and facilitate follow-up care. These logistical aspects contribute just as significantly as the clinical elements to the success of the programs.

Furthermore, ethical considerations are interwoven into every facet of newborn screening. Parental informed consent is crucial. While screening is generally considered a public health imperative, parents should have the right to decline testing. However, this decision-making process is frequently complex, requiring careful consideration of potential benefits and risks. Providing comprehensive information about the tested conditions, their associated risks and benefits, and available treatments empowers parents to make truly informed choices. This informed consent is ethically critical in newborn screening.

Beyond individual choices, broader societal considerations are paramount. Ensuring equitable access to screening, irrespective of geographic location or socioeconomic background, is vital. Public health authorities must invest in comprehensive infrastructure and resources to ensure widespread access. Furthermore, the cost-effectiveness of screening programs must be regularly evaluated. The cost-benefit analysis should weigh the cost of screening against the potential long-term health and economic benefits of early intervention. This involves weighing the costs of the test against the long-term healthcare costs of treating the disease. This is a continuing need.

Finally, the role of genetic counselors and other healthcare professionals cannot be overstated. They provide crucial support throughout the process. Genetic counselors play a critical role in explaining test results to parents, counseling them about the implications of positive screenings, and guiding them through subsequent medical management. They translate complex genetic information into understandable terms, providing both emotional and informational support. Such support is invaluable for families facing potentially life-altering news. The role of the genetic counselor in communicating with the parents cannot be underestimated. Their guidance and expertise are vital for families. Effective communication and comprehensive support systems are therefore essential for the success and ethical conduct of newborn screening initiatives.

4.5 Direct-to-Consumer Testing

Direct-to-consumer (DTC) genetic tests offer accessible insights into your ancestry, predispositions to certain diseases, and even carrier status for various genetic conditions. These tests, readily available online or at retail stores, bypass the traditional physician-ordered testing pathway, leading to both advantages and considerable drawbacks. Their ease of access and affordability make them incredibly appealing, promising a glimpse into your genetic makeup without the hassle of doctor's appointments and potentially expensive medical tests. However, it's crucial to understand the limitations and potential pitfalls before embarking on this seemingly straightforward journey of self-discovery.

One of the primary concerns surrounding DTC genetic testing is the accuracy and interpretation of the results. Many tests focus on a limited number of genetic markers, potentially missing important information. Furthermore, the interpretation of these results often lacks the nuanced context provided by a genetic counselor or physician. This can lead to misunderstandings, anxiety, and potentially inappropriate healthcare decisions based on incomplete or misinterpreted data. A simple example is a predisposition to a disease: the test may indicate a higher risk, but this does not equate to a diagnosis. The probability, severity, and timeline for development of that condition still needs significant context from a qualified professional. It's not just about the raw numbers; it's about how those numbers translate into real-world health implications.

The lack of regulation and standardization in the DTC genetic testing market is a significant concern. Unlike tests ordered by a healthcare provider, many DTC tests aren't subject to the same rigorous quality control and validation procedures. This lack of oversight raises questions about the accuracy and reliability of the results. Furthermore, the lack of consistent standards in reporting and interpreting results makes it difficult to compare findings across different companies, further complicating the process. Consider the diverse methods used to determine your ancestry; the difference in algorithms and data sets can lead to widely varying conclusions from test to test. A thorough understanding of the limitations of the technology and the company's methodology is therefore paramount.

Beyond the analytical challenges, the emotional and psychological ramifications of DTC genetic testing warrant careful consideration. Receiving unexpected or difficult-to-interpret information about your health or ancestry can be emotionally jarring. For instance, discovering a predisposition to a serious illness can cause significant anxiety and distress, even if the risk is relatively low. Similarly, unexpected results regarding ancestry can challenge long-held beliefs and family narratives. These tests can have a profound emotional impact. Support mechanisms, including access to genetic counseling, are critical in mitigating the potential for negative psychological consequences, ensuring users understand the implications and are able to process the information effectively and healthily.

The implications of DTC genetic testing extend beyond the individual to broader societal concerns. Data privacy is a major issue, as many companies collect and store extensive genetic information. The potential for misuse of this data, including for insurance discrimination or marketing purposes, raises significant ethical concerns. Regulation is struggling to keep pace with the rapid advancements in technology and the growing popularity of DTC testing. Striking a balance between protecting individual privacy and fostering innovation in this space is a significant challenge. The potential for misuse, both intentionally and unintentionally, should always be in the forefront of any decision regarding participation in this kind of testing.

In conclusion, while DTC genetic testing presents an exciting opportunity for individuals to explore their genetic makeup, it's crucial to approach it with a healthy dose of skepticism and caution. The limitations in accuracy, interpretation, and regulation, combined with the potential for emotional and psychological distress, highlight the need for careful consideration before utilizing these tests. While the promise of personalized health and ancestry insights is appealing, it's vital to understand the context, consult with healthcare professionals, and critically assess the limitations before making informed decisions based on the results received. Always remember that these tests provide only a snapshot of your genetic makeup, and professional medical guidance is essential for interpreting results and making health decisions.

5 Ethical Implications of Genetic Engineering

Genetic engineering offers unprecedented power to alter life itself, raising complex ethical questions that demand careful consideration. We've seen remarkable advancements, from CRISPR's precise gene editing to gene therapies offering hope for previously incurable diseases. However, this progress comes with significant moral and societal challenges. The potential for misuse, the unpredictable consequences, and the inherent inequities these technologies could exacerbate require thoughtful discussion and robust regulatory frameworks. Failing to address these issues head-on risks compromising the very benefits genetic engineering promises.

One major concern revolves around genetic discrimination. Imagine a future where insurance companies deny coverage or employers refuse employment based on an individual's genetic predisposition to certain diseases. This scenario, while chilling, isn't purely hypothetical. Existing laws in many countries offer some protection, but the rapid advancement of genetic testing technologies outpaces our ability to create comprehensive and truly effective safeguards. We need stronger legal protections, coupled with public education initiatives to combat prejudice and ensure fair access to opportunities regardless of genetic makeup. Stronger international cooperation is also needed to establish common standards and prevent a "genetic arms race".

Germline editing presents arguably the most profound ethical dilemma. Unlike somatic cell gene editing, which affects only the individual, germline editing alters the genes passed down to future generations. This has the potential to introduce unforeseen and irreversible changes into the human gene pool. While proponents argue it could eradicate inherited diseases, critics express deep concerns about unintended consequences and the potential for eugenics, the selective breeding of humans based on desirable traits. The slippery slope argument holds strong weight here; where do we draw the line between disease prevention and enhancement? Who decides which traits are deemed "desirable"? Open and inclusive public discourse, involving scientists, ethicists, policymakers, and the public, is essential to navigate this intricate landscape. International consensus and strict ethical guidelines must precede any widespread application of this technology.

Beyond germline editing, broader access to genetic technologies poses a challenge. The cost of genetic testing and therapies often makes them inaccessible to many, particularly in developing countries. This creates a concerning disparity, exacerbating existing health inequities. Furthermore, ensuring equitable access requires not only financial considerations but also addressing disparities in education and healthcare infrastructure. The potential for these technologies to worsen existing global inequalities demands proactive efforts to guarantee equitable distribution and access. This could involve international collaborations, philanthropic initiatives, and the development of cost-effective technologies. Ultimately, ensuring that genetic engineering benefits all of humanity, not just the privileged few, is crucial.

The responsibility for navigating these complex ethical issues rests on multiple shoulders. Scientists must adhere to the highest ethical standards, conducting thorough research and openly communicating their findings. Policymakers must develop and implement regulations that balance the promise of genetic engineering with its potential risks. And the public needs to engage in informed discussions, shaping policy and promoting responsible innovation. Without a collaborative and ethically guided approach, the immense potential of genetic engineering risks being overshadowed by its unforeseen and potentially devastating consequences. The future of genetic engineering will be determined not only by technological advancements but also by our ability to address the profound ethical challenges it presents. Only through responsible innovation, guided by careful ethical considerations and a commitment to equity, can we harness the transformative power of genetic engineering for the benefit of all humankind.

5.1 Genetic Discrimination

Genetic discrimination, the unfair or prejudicial treatment of individuals based on their genetic information, is a significant ethical concern arising from advancements in genetic testing and understanding. It's not merely a hypothetical threat; real-world instances demonstrate the potential for this bias to manifest in employment, insurance, and even social interactions. Imagine an individual denied health insurance because a predictive test suggests a heightened risk of a future illness, regardless of their current health status. Or consider the job applicant passed over due to a genetic predisposition to a certain condition, even if it's not currently affecting their work capabilities. These scenarios, sadly, are not far-fetched.

The potential for misuse of genetic information is immense. Employers could utilize genetic data to discriminate against workers deemed to be high-risk, leading to job losses and economic hardship. Insurance companies might increase premiums or refuse coverage altogether based on genetic predispositions, creating significant financial burdens for families. Such actions violate fundamental principles of fairness and equity, exacerbating existing health and economic disparities. The fear of this discrimination can also deter individuals from undergoing crucial genetic testing, hindering preventative care and early intervention. This hesitancy can have profound consequences for personal health and overall societal well-being.

Legislation attempts to mitigate this risk vary widely across different jurisdictions. Some countries and states have enacted laws prohibiting genetic discrimination in employment and insurance, protecting individuals from adverse consequences based solely on their genetic makeup. However, these laws aren't always consistently enforced, and loopholes frequently exist. Furthermore, even where laws exist, proving genetic discrimination can be incredibly challenging. Often, there is indirect discrimination; it's difficult to establish a direct causal link between a negative outcome and discriminatory actions based on genetic information, resulting in a lack of legal redress for those affected. These gaps and challenges highlight the need for stricter laws and robust enforcement mechanisms.

Beyond formal legal frameworks, societal awareness and proactive measures are crucial. Educating the public about genetic discrimination and its implications is paramount. This includes clarifying misconceptions about genetic predispositions and emphasizing that having a genetic risk factor does not equate to a guaranteed outcome. Furthermore, initiatives should promote responsible data handling practices within the healthcare and insurance sectors, and robust data privacy regulations are essential to safeguard genetic information from misuse. Companies and institutions should be trained in ethical handling of genetic data, emphasizing the importance of confidentiality and non-discrimination policies. Such policies must extend beyond simple compliance and incorporate a commitment to fairness and equity in all decision-making processes that involve genetic information.

A multi-pronged approach is necessary to effectively combat genetic discrimination. This requires a combination of strong legal frameworks, enhanced enforcement mechanisms, broader public education campaigns, and a strong commitment from institutions to prioritize ethical considerations in the handling of genetic data. Without these concerted efforts, the benefits of personalized medicine and genetic advancements could be overshadowed by the profound inequalities created by genetic discrimination, perpetuating and widening existing societal divides. This chapter highlights the critical importance of balancing the potential benefits of genetic technologies with the imperative of safeguarding against its discriminatory potential, ensuring that advances in genetic knowledge promote health equity and societal well-being for everyone. Ignoring these ethical considerations risks a future where the promise of personalized medicine remains out of reach for many. We must actively work to shape a future where genetic knowledge is a tool for progress, not a weapon of prejudice.

5.2 Germline Editing Concerns

Germline editing, the modification of genes within reproductive cells, presents a unique set of challenges absent in somatic cell editing. Unlike somatic changes, which affect only the individual, germline alterations are heritable, impacting future generations. This introduces profound ethical concerns that demand careful consideration. The potential for unforeseen consequences, both intended and unintended, looms large. We are, in essence, rewriting the human genetic code, a responsibility that carries significant weight.

The long-term effects of germline modifications are largely unknown. While current technologies allow for precise targeting of specific genes, the intricate interplay of genes within the human genome means that even subtle changes could have unpredictable consequences across multiple generations. Off-target effects, the unintended modification of other genes, remain a significant risk, potentially leading to unforeseen health problems. We must proceed with extreme caution, prioritizing thorough research and rigorous testing before any widespread application.

Furthermore, the equitable access to germline editing technologies is a serious concern. The high costs associated with these procedures will likely create a disparity, where the wealthy gain access to enhancements and disease prevention, while marginalized communities are left behind. This could exacerbate existing societal inequalities, potentially creating a genetic divide. This inequitable access presents a profound ethical challenge, emphasizing the need for global discussions and policies to ensure fair and just distribution.

Beyond the practical challenges lie complex philosophical and societal considerations. Germline editing raises questions about the very definition of humanity. Is it ethical to intervene in the natural course of human evolution? What constitutes a justifiable reason for altering the human germline? Who decides which traits should be enhanced or eliminated? These are not merely scientific questions, but also deeply moral ones, requiring input from diverse fields including ethics, philosophy, and law, along with robust public dialogue.

The potential for misuse of germline editing technologies is another significant concern. The power to manipulate human heredity could be exploited for non-therapeutic purposes, such as creating "designer babies" with specific traits. This raises the specter of eugenics, a dark chapter in human history marked by discriminatory and harmful practices. Safeguards and regulations are necessary to prevent such misuse, ensuring that germline editing remains within ethical and socially responsible boundaries.

Consider the case of a hypothetical scenario: parents opt for germline editing to eliminate a hereditary disease risk in their future children. While seemingly noble, this decision could have unintended consequences for future generations. The edited gene could interact unexpectedly with other genes, leading to unforeseen health problems. Moreover, the elimination of certain genes might inadvertently reduce human genetic diversity, potentially making future generations more vulnerable to environmental changes or emerging diseases. These are vital considerations that need careful and extensive study, with the help of experts in various fields.

Therefore, before embracing germline editing on a wider scale, rigorous ethical frameworks need to be established. These frameworks should consider not only the short-term benefits but also the potential long-term consequences for individuals, families, and society as a whole. Public engagement and open dialogue are essential in shaping policies that ensure responsible innovation and minimize risks. The challenge is not merely to develop the technology but also to develop the societal wisdom to guide its use wisely and ethically. The development of sophisticated tools requires thoughtful ethical oversight, and must ensure fairness, equity and justice in the process.

In conclusion, germline editing holds immense potential for preventing inherited diseases and enhancing human health. However, the significant ethical challenges inherent in altering the human germline demand careful consideration. Thorough research, robust ethical guidelines, equitable access, and informed public discourse are essential to navigate the complex implications of this powerful technology and ensure its responsible application for the benefit of humanity. The future of germline editing depends not only on scientific advancements but also on the ethical wisdom with which we guide its development and implementation. This is not just a scientific endeavor, but a moral imperative. Moving forward, collaboration between scientists, ethicists, policymakers, and the public is paramount to ensuring a future where this technology serves humanity's best interests. Failure to address these concerns could lead to unforeseen consequences, potentially harming the very people it intends to help.

6 Future of Genetic Engineering

Predicting the future is always a gamble, but looking at the current trajectory of genetic engineering, several exciting, and potentially disruptive, trends emerge. Gene drives, for example, represent a paradigm shift in genetic manipulation. These engineered genetic elements can spread rapidly through populations, offering the potential to eradicate disease vectors like mosquitoes carrying malaria. However, the potential for unintended ecological consequences necessitates careful consideration and robust risk assessment frameworks. Unforeseen disruptions to ecosystems are a serious concern that needs rigorous study and open dialogue before widespread deployment. We must prioritize responsible innovation in this area.

Synthetic biology, another rapidly advancing field, promises to revolutionize various sectors. Imagine designing microbes to produce sustainable biofuels, or engineering plants that require less water and fertilizer. The possibilities are virtually limitless, yet we must also address potential biosecurity risks. The creation of novel biological systems carries inherent uncertainties, and establishing appropriate safeguards is paramount to prevent malicious use or accidental release. International collaborations and stringent regulatory frameworks are crucial.

Beyond technological advancements, the future of genetic engineering hinges on ethical considerations. Developing robust ethical frameworks is not merely an academic exercise; it's a necessity for navigating the complex moral dilemmas this technology presents. Public discourse and education are essential to shape societal values and expectations concerning the use of gene editing. We need to foster informed conversations about the acceptable limits of genetic modification. Transparency in research and development, alongside meaningful public engagement, is crucial to build trust and ensure equitable access to these transformative technologies. We need to consider not just the scientific possibilities but the societal impact to ensure that ethical considerations guide progress, rather than lagging behind it.

Genetic engineering's future impact on healthcare is profoundly transformative. We can anticipate personalized medicine becoming increasingly prevalent, with treatments tailored to an individual's unique genetic makeup. This means more effective therapies, fewer side effects, and potentially the prevention of diseases before they even manifest. However, this progress also raises concerns about accessibility and equity. Ensuring that these advances benefit everyone, regardless of socioeconomic status, is a critical challenge. Affordable access to personalized genetic medicine must be a central goal, alongside addressing potential biases in its application.

The ethical frameworks guiding genetic engineering must evolve alongside the technology itself. This means establishing clear guidelines for gene editing in humans, considering the potential impact on future generations and addressing the social implications of genetic enhancement. The principles of autonomy, beneficence, nonmaleficence, and justice must be central to any ethical framework, ensuring that genetic technologies are used responsibly and ethically. International cooperation and the establishment of independent oversight bodies are essential for enforcing ethical guidelines and promoting responsible innovation. We must be proactive in anticipation of the coming transformations to create ethical frameworks adapted to ongoing scientific breakthroughs.

Looking ahead, the convergence of genetic engineering with other fields like artificial intelligence and nanotechnology promises further breakthroughs. This convergence could lead to powerful new tools for disease diagnosis, treatment, and prevention. However, this convergence also necessitates a deeper exploration of the ethical, social, and environmental implications. The challenges will be substantial, requiring a multidisciplinary approach involving scientists, ethicists, policymakers, and the public. Understanding the potential impacts of this converging technology is vital for making informed decisions about its development and implementation. Proactive assessment of potential issues and the collaborative efforts of diverse experts are crucial for navigating the uncharted territory.

The future of genetic engineering is not predetermined; it is being shaped by the choices we make today. By fostering open dialogue, promoting responsible innovation, and prioritizing ethical considerations, we can harness the transformative power of this technology to improve human health and well-being, while mitigating potential risks. This requires a commitment to ongoing dialogue, the constant reevaluation of ethical guidelines, and the development of robust regulatory frameworks, all underpinned by a deep understanding of the science and societal implications of this rapidly advancing field. Ultimately, the future of genetic engineering depends on our collective wisdom and commitment to responsible stewardship. The path forward requires a shared vision of a future where genetic technologies benefit all humankind.

6.1 Gene Drives

Gene drives represent a revolutionary advancement in genetic engineering, capable of altering the genetic makeup of entire populations. Unlike traditional genetic modifications, which rely on natural selection, gene drives ensure near-complete inheritance of a modified gene within a population. This is achieved through genetic elements that bias inheritance, effectively making the modified gene nearly impossible to eliminate. Imagine a gene drive targeting a disease-carrying mosquito population; the modified gene, preventing disease transmission, would rapidly spread through the population, dramatically reducing the incidence of the disease. The implications are vast, potentially offering solutions to numerous global challenges.
However, the power of gene drives necessitates cautious consideration. Their potential to irreversibly alter ecosystems necessitates rigorous risk assessment and containment strategies. Unforeseen consequences, ecological imbalances, and the potential for unintended spread to non-target species are legitimate concerns requiring careful investigation and mitigation. The development of sophisticated modeling techniques to predict the potential impacts of gene drives on complex ecosystems is crucial. We need tools to accurately simulate the spread of the altered gene, account for population dynamics, and forecast potential environmental consequences.

Consider the ethical dilemmas. The ability to manipulate entire populations raises serious ethical questions, especially concerning the potential for unintended environmental damage or unforeseen health impacts. The lack of fully developed methodologies for reversing or containing a gene drive's effects creates concerns about unintended consequences that could be difficult, if not impossible, to undo. Transparency and public engagement are, therefore, paramount in the development and deployment of this technology. Open discussions involving scientists, policymakers, and the public are essential to establish clear guidelines and regulations.

The potential applications of gene drives extend beyond disease control. They could be used to control invasive species, combat agricultural pests, and even enhance biodiversity. However, this potential must be balanced against the risks. A thorough understanding of the target ecosystem and the potential for unintended consequences is essential. Before deployment, there must be a robust and comprehensive risk assessment procedure involving multiple scientific disciplines, including ecology, genetics, and environmental science. Independent expert review of these assessments should be mandatory.

Responsible development necessitates a stringent regulatory framework. This framework must address aspects like pre-release testing, containment measures, and monitoring strategies to ensure that gene drives are deployed responsibly. It's not merely about prohibiting or permitting, but about establishing a clear and transparent system to carefully assess and manage the risks. This framework should be international in scope, considering the potential for gene drives to cross geographical boundaries. Collaboration between nations is essential to create consistent and effective regulations, preventing potential abuses or unintended global consequences.

Moreover, ethical considerations must be at the forefront of any gene drive project. This extends beyond environmental impacts to the broader social and societal implications. What role should public input play in the decision-making process? How can we ensure equity and avoid potential exploitation or harm to marginalized communities? These ethical questions require careful consideration and proactive discussion, engaging diverse perspectives from various stakeholders. The development of guiding ethical principles will be crucial in shaping responsible research and deployment of gene drive technologies.

Further research is essential to improve the precision and safety of gene drives. This includes advancements in gene-editing technologies, developing sophisticated monitoring systems, and improving predictive models for assessing the ecological consequences of releasing modified organisms. In addition to these scientific efforts, extensive public education is needed to foster a broad understanding of gene drives, their potential benefits, and associated risks. It is only through informed public discourse and rigorous scientific investigation that the full potential of gene drives can be responsibly harnessed, minimizing risks and maximizing societal benefit.

In conclusion, gene drives offer tremendous potential for addressing various global challenges but present significant ethical and ecological considerations. Responsible development and deployment necessitate a multidisciplinary approach, involving scientists, ethicists, policymakers, and the public. A comprehensive regulatory framework, coupled with rigorous risk assessment and ongoing research, is crucial for ensuring that this revolutionary technology is utilized responsibly, ultimately benefiting humanity while mitigating potential negative consequences. The future of gene drives depends on a balance between innovation and precaution, a delicate dance between potential and responsibility.

6.2 Synthetic Biology

Synthetic biology represents a groundbreaking frontier in genetic engineering, pushing the boundaries of what's possible by designing and constructing new biological parts, devices, and systems. Imagine creating entirely new organisms with tailored characteristics, or engineering existing ones to perform novel functions – that's the power of synthetic biology. It's a field brimming with potential, from developing sustainable biofuels and biodegradable plastics to creating novel medicines and advanced materials. However, navigating this field requires a deep understanding of its capabilities and limitations. We can achieve impressive feats with this technology. For example, researchers have already successfully synthesized entire viral genomes and even bacterial chromosomes, demonstrating the feasibility of creating life from scratch. This opens up unprecedented opportunities for studying fundamental biological processes and engineering organisms with specific functionalities. The implications are truly transformative, paving the way for advancements in various sectors, from medicine to manufacturing.
However, this technological advancement also brings about certain challenges. The design and construction of complex biological systems is incredibly intricate and demands precise control over numerous interacting components. Predicting the behavior of these engineered systems can be difficult, and unintended consequences are a

possibility. Rigorous testing and careful evaluation are critical to ensure the safety and effectiveness of synthetic biology applications.
One of the most promising applications of synthetic biology lies in the realm of medicine. We could potentially design microorganisms to produce novel therapeutic proteins, develop advanced diagnostic tools, or even engineer cells to fight diseases within the body. Imagine a future where customized therapies are created at the molecular level, perfectly tailored to individual patients' genetic makeup. This precision medicine approach could revolutionize healthcare, leading to more effective and personalized treatments.

Yet, ethical considerations are paramount. The creation of entirely new life forms raises significant ethical concerns regarding their potential impact on the environment and human society. Furthermore, the misuse of synthetic biology technologies for malicious purposes, such as creating bioweapons, presents a substantial risk. Therefore, robust ethical guidelines and regulatory frameworks are absolutely essential to guide responsible development and prevent potential harms.

Developing synthetic biology solutions requires interdisciplinary collaboration. Biologists, engineers, chemists, and computer scientists must work together to overcome the technical challenges and ensure the safe and responsible use of this powerful technology. This necessitates the creation of strong partnerships between academia, industry, and government, fostering open communication and collaborative efforts.

The educational aspect is crucial for the responsible use and advancement of this field. Educating future scientists, engineers, and policymakers on the capabilities and ethical implications of synthetic biology is critical for navigating the complex landscape of this evolving field. We need to cultivate a responsible and innovative scientific community, and this starts with education.

The future of synthetic biology is brimming with potential. It holds the key to tackling some of humanity's most pressing challenges, such as climate change, disease, and food security. However, this potential must be harnessed responsibly. By carefully considering the ethical implications, promoting open collaboration, and establishing strong regulatory frameworks, we can ensure that the promise of synthetic biology is realized for the benefit of all. The future of synthetic biology depends on our collective wisdom and commitment to responsible innovation.

Ultimately, synthetic biology represents a powerful tool with both immense potential and significant risks. The careful and ethical application of this technology will be crucial in shaping a future where its benefits are maximized while potential harms are minimized. We stand at the cusp of a new era in biology, and the choices we make today will determine the course of this remarkable field.

6.3 Ethical Frameworks

Utilitarianism, a cornerstone of ethical decision-making, emphasizes maximizing overall well-being. In the context of genetic engineering, this translates to weighing the potential benefits—like eradicating inherited diseases—against potential harms, such as unforeseen consequences or exacerbating existing inequalities. A utilitarian approach necessitates a rigorous cost-benefit analysis, considering the potential impact on individuals, communities, and future generations. This requires careful consideration of factors like accessibility, affordability, and the potential for misuse. We must ensure that the benefits are broadly distributed and that the potential for harm is minimized through stringent regulation and oversight. A truly utilitarian approach demands continuous monitoring and adaptation of policies to reflect evolving scientific understanding and societal values. Implementing such a framework requires robust public discourse and transparent decision-making processes.

Deontology, on the other hand, prioritizes moral duties and rules regardless of outcome. A deontological framework in genetic engineering would focus on inherent rights and responsibilities. This approach might prioritize respecting individual autonomy—ensuring informed consent in all genetic interventions—and avoiding actions that could be deemed intrinsically wrong, such as manipulating human germline cells to enhance traits rather than prevent disease. Strict adherence to established ethical guidelines and regulations would be paramount. The focus shifts from maximizing overall good to upholding fundamental moral principles. This could lead to restrictions on certain genetic engineering applications deemed ethically problematic, even if they might offer significant benefits. The challenge lies in defining and prioritizing these moral duties in a rapidly evolving field with uncertain long-term consequences.

Virtue ethics, a third significant framework, centers on character and moral excellence. Applied to genetic engineering, this approach would emphasize the virtues of scientists, policymakers, and the public. It focuses on cultivating traits like prudence, compassion, and justice in the development and application of these technologies. This approach would necessitate robust education and training programs to instill ethical awareness and responsible behavior in all stakeholders. It would also emphasize transparency and accountability in research, development, and implementation. The goal is not just to develop effective technologies but to do so in a manner that reflects the highest moral standards. This involves ongoing reflection and self-assessment, ensuring that actions align with a commitment to ethical excellence. Fostering these virtues requires a societal commitment to ethical education and a culture of responsible innovation. Integrating these three frameworks—utilitarianism, deontology, and virtue ethics—provides a robust and comprehensive ethical approach to genetic engineering. A truly holistic approach acknowledges the complex interplay between maximizing overall well-being, upholding moral duties, and fostering virtuous behavior. It necessitates ongoing dialogue and collaboration between scientists, ethicists, policymakers, and the public. A dynamic ethical framework—one that adapts to new knowledge and societal values—is essential for navigating the complex ethical challenges presented by genetic engineering. Only through a thoughtful and inclusive approach can we harness the immense potential of this technology while mitigating its risks and ensuring its benefits are shared equitably. The development of such a comprehensive ethical framework is a continuous process, requiring ongoing reflection and adaptation to the ever-evolving landscape of genetic engineering and its societal implications. This necessitates a multidisciplinary approach, drawing upon expertise from diverse fields, fostering open dialogue, and ensuring that ethical considerations are integrated into every stage of the research, development, and application of genetic engineering technologies. Only then can we hope to harness the full potential of this field while mitigating the risks to humanity and the environment.

The development and implementation of these ethical frameworks must be transparent and participatory. Open public discourse is crucial in shaping the ethical boundaries of genetic engineering. This demands inclusive dialogues engaging all relevant stakeholders—scientists, healthcare professionals, policymakers, religious leaders, and the public—to ensure that ethical considerations reflect the values and concerns of society as a whole. Engaging with diverse perspectives and fostering critical discussions are vital in navigating the complex ethical landscape of this rapidly evolving field. Furthermore, ethical frameworks must not only guide research and development but also inform the application and regulation of genetic technologies, ensuring responsible innovation and equitable access. Continuous evaluation and refinement of these frameworks are paramount in adapting to scientific advances and shifting societal values, thereby fostering a responsible and ethically sound trajectory for genetic engineering. The creation of a robust and dynamic ethical framework is essential for guiding the responsible and equitable development and use of genetic engineering.

7 Conclusion: The Genetic Revolution

The genetic revolution is undeniably transforming our world. Its impact extends far beyond the laboratory, weaving itself into the fabric of society, reshaping healthcare, agriculture, and even our understanding of life itself. This profound shift necessitates a careful consideration of its implications, prompting us to grapple with complex ethical questions and strategize for a future where genetic technologies are used responsibly. We stand at a pivotal moment, where the choices we make today will define the trajectory of tomorrow.

Understanding the full potential of gene editing tools like CRISPR-Cas9 requires continued research. While we've seen remarkable progress in treating genetic diseases, there are still significant hurdles to overcome. Further investigation into off-target effects, long-term consequences, and the development of safer and more efficient delivery systems are crucial for realizing the full therapeutic potential of these technologies. This necessitates collaboration between researchers, clinicians, and regulatory bodies worldwide, ensuring the responsible application of gene editing across diverse populations. We must approach this advancement with a deep understanding of the potential risks alongside the anticipated rewards.

Beyond therapeutic applications, genetic engineering holds immense potential in addressing global challenges. Consider the possibilities in agriculture, where gene editing could be harnessed to develop crops resistant to pests, diseases, and harsh environmental conditions. Imagine the impact on food security, particularly in regions struggling with famine or climate change-induced agricultural losses. However, the release of genetically modified organisms into the environment requires rigorous risk assessment and careful consideration of unintended ecological consequences. Open dialogue and robust regulatory frameworks are crucial to balancing the benefits of such innovations with the potential environmental risks they might pose. Societal implications are vast and varied. Genetic testing, once a niche field, is becoming increasingly accessible and affordable, raising questions about genetic privacy, discrimination, and the potential for misuse of personal genetic information. Establishing comprehensive data protection laws and promoting public awareness of genetic privacy rights are paramount. We must ensure that advancements in genetic technology benefit all members of society equitably, without exacerbating existing inequalities. Access to these powerful tools must be fair and just; this isn't just a scientific endeavor but a societal responsibility.

The ethical considerations surrounding germline editing remain a topic of intense debate. The prospect of altering the human germline, affecting future generations, raises profound ethical concerns. International collaborations are required to establish global guidelines and regulations to guide the responsible development and application of these powerful technologies. Furthermore, continuous public discourse is essential to build societal consensus on the ethical boundaries of germline editing. Open and transparent discussions involving scientists, ethicists, policymakers, and the public are crucial for establishing ethical norms and safeguarding human dignity.

Moving forward, we need a multidisciplinary approach involving scientists, ethicists, policymakers, and the public to navigate the complexities of this technological revolution. Educational initiatives are crucial to fostering public understanding of genetics and its implications. This will empower individuals to make informed decisions about genetic testing and other genetic technologies. Moreover, responsible innovation requires a commitment to transparency, public engagement, and the development of ethical frameworks that guide the research, development, and application of genetic engineering technologies. This inclusive process is vital for ensuring that these powerful tools are used to benefit humanity, while minimizing risks and maximizing potential positive impact.

The journey into the era of genetic engineering is far from over. It's a continuing exploration, one that necessitates a commitment to ongoing research, ethical reflection, and responsible innovation. We are at a crucial juncture, shaping the future of healthcare, agriculture, and our very understanding of life itself. The choices we make today will determine whether we harness the transformative power of genetics to improve human lives or succumb to its potential risks. The future of genetic engineering hinges upon our collective wisdom, foresight, and unwavering commitment to responsible innovation. The challenges are substantial; the potential rewards are immense. Only through careful consideration and collaborative efforts can we ensure that the genetic revolution truly benefits all of humanity.

7.1 Societal Impacts

The democratization of genetic information presents both unprecedented opportunities and significant challenges. Access to affordable and readily available genetic testing, once a privilege, is increasingly becoming a reality. This empowers individuals to make informed decisions about their health, family planning, and lifestyle choices. However, this accessibility also raises concerns about potential misuse of this information by employers, insurance companies, or even educational institutions. Regulations are crucial to prevent genetic discrimination and ensure equitable access to this powerful technology for everyone, regardless of socioeconomic status. The potential for increased health disparities based on access must be mitigated actively. We need to create a robust ethical framework guiding responsible innovation to ensure benefits are shared widely, without exacerbating existing inequalities. This necessitates public education initiatives and transparent policies ensuring that the benefits of genetic technology aren't confined to a privileged few.

Genetic engineering's societal impact extends far beyond individual healthcare. Consider agricultural advancements: gene editing holds immense potential for improving crop yields, enhancing nutritional value, and increasing resistance to pests and diseases. This could be pivotal in addressing global food security challenges, particularly in regions facing climate change impacts. However, the release of genetically modified organisms into the environment raises questions about unintended ecological consequences. Rigorous testing and regulatory oversight are essential to minimize risks and ensure the responsible development and deployment of genetically modified crops. The long-term effects on biodiversity and ecosystem stability must be carefully considered before widespread adoption. International collaborations and consistent regulations are necessary to navigate the complexities of this area.

The ethical considerations surrounding germline editing—altering the genes passed on to future generations—are particularly complex and demanding. While offering the potential to eradicate inherited diseases, it raises profound questions about altering the human gene pool and the potential for unforeseen consequences. We must engage in open and inclusive discussions about the limits of human intervention in reproduction, acknowledging the potential for unintended consequences and the necessity of establishing stringent ethical guidelines. Societal consensus is crucial to guide the development and application of this technology, involving diverse stakeholders, from scientists and ethicists to policymakers and the public at large.

Furthermore, the use of genetic information in criminal investigations raises complex legal and ethical dilemmas. DNA evidence plays a crucial role in solving crimes, but questions of privacy, data security, and the potential for misinterpretation or misuse of genetic data need careful attention. Protecting individuals' privacy and ensuring the integrity of forensic genetic analysis are paramount. We need clear guidelines regarding data storage, access, and analysis to prevent wrongful convictions and maintain public trust in the legal system. Strict regulations, combined with ongoing scientific advancements in forensic genetics, are vital to ensure the ethical and responsible use of genetic data within the criminal justice system.

Beyond these specific areas, the integration of genetic engineering into society requires addressing various social and cultural implications. Public understanding and acceptance of genetic technologies are vital for their successful implementation. This requires effective communication by scientists and policymakers, and inclusive public engagement fostering trust and addressing concerns. Educational initiatives to increase genetic literacy among the public are crucial, enabling informed discussions and decision-making. We must work toward ensuring that the benefits of genetic advancements are enjoyed by all members of society, while minimizing potential risks and negative consequences. This involves promoting equitable access to information and resources, and tackling health disparities stemming from genetic testing and treatment.

Addressing the societal impacts of genetic engineering requires a multi-faceted approach, including rigorous scientific research, robust ethical frameworks, and transparent public engagement. Only through a collaborative effort involving scientists, ethicists, policymakers, and the public can we harness the power of genetic technology responsibly, ensuring that its benefits are widely shared, while mitigating potential risks. This is not merely a scientific endeavor; it's a societal challenge requiring a comprehensive and proactive approach. The future of genetic engineering depends upon our collective wisdom and commitment to responsible innovation. The path forward necessitates a proactive approach, balancing scientific advancement with ethical considerations and societal needs. A constant dialogue and collaborative effort are essential to navigating the complexities and opportunities that lie ahead.

7.2 Future Research

The field of genomics is rapidly expanding, presenting numerous avenues for future research. One particularly promising area lies in improving the accuracy and efficiency of gene editing tools like CRISPR-Cas9. Current limitations include off-target effects—unintended modifications to the genome—and challenges in delivering the editing machinery to specific cells within the body. Future research should focus on developing more precise and targeted gene editing techniques, potentially involving advanced delivery systems like nanoparticles or engineered viruses. Simultaneously, sophisticated computational modeling and AI-assisted design could significantly enhance the precision and predictability of gene editing, minimizing the risk of unintended consequences. The development of robust in silico testing protocols could also significantly reduce reliance on extensive and costly in vivo experiments.

Understanding the complex interplay between genetics and the environment remains a critical area for investigation. While our understanding of genetic predisposition to diseases is growing, we still lack a comprehensive grasp of how environmental factors, including lifestyle choices, diet, and exposure to toxins, interact with our genes to influence disease risk and progression. Epigenomics, the study of heritable changes in gene expression without altering the DNA sequence itself, holds immense promise in this context. Further research into epigenetic modifications, their stability, and their reversibility could unlock new therapeutic strategies for preventing and treating a wide range of diseases. Large-scale epidemiological studies combining genomic data with detailed environmental and lifestyle information are essential for advancing our knowledge in this area, requiring international collaboration and robust data sharing protocols.

Another key area requiring further exploration is the development of novel gene therapy strategies. While gene therapy has shown remarkable progress, challenges remain in achieving long-term efficacy and reducing the risk of immune responses. The development of more effective gene delivery systems, potentially involving non-viral vectors or sophisticated viral vector engineering, is critical. Additionally, research into gene editing strategies that can specifically target disease-causing genes without affecting other parts of the genome is crucial. Moreover, studying the long-term effects of gene therapies and developing robust monitoring systems are necessary to ensure patient safety and long-term health. This requires a multifaceted approach, integrating genetic, immunological, and clinical data.

Personalized medicine, though rapidly advancing, requires continued research to fully realize its potential. The development of more accurate and efficient methods for predicting individual responses to medications based on their genetic makeup is essential. This necessitates the development of sophisticated algorithms and machine learning models capable of integrating diverse datasets, including genomic data, clinical information, and lifestyle factors. Similarly, research into the genetic basis of drug resistance and the development of targeted therapies tailored to individual cancer genomes needs to continue. The integration of advanced imaging techniques with genomic information could revolutionize the early detection and precise treatment of various cancers. Realizing the potential of personalized medicine requires not just technical innovation, but also addressing ethical considerations related to data privacy and equitable access to these advanced technologies.

Beyond these specific areas, future research needs to address the ethical, social, and legal implications of advanced genetic technologies.

The development of ethical guidelines and regulatory frameworks to govern the use of gene editing and other powerful genetic technologies is crucial. Furthermore, public education and engagement are vital to foster informed decision-making and ensure the responsible application of these technologies. Broader conversations about genetic literacy and responsible innovation need to engage researchers, policymakers, and the general public to forge responsible and ethical applications. These discussions should center around the potential benefits, risks, and societal implications of advancements in this field. This requires a collaborative, interdisciplinary effort, engaging diverse perspectives and ensuring equitable outcomes.

7.3 Responsible Innovation

The transformative potential of genetic engineering is undeniable; it holds the key to treating previously incurable diseases, enhancing crop yields, and even potentially eradicating inherited conditions. However, this power comes with a significant responsibility. The unchecked advancement of these technologies risks exacerbating existing inequalities, creating unforeseen ecological consequences, and raising profound ethical dilemmas. Responsible innovation, therefore, isn't merely a desirable outcome; it's an absolute necessity.
A key aspect of responsible innovation is prioritizing transparency and open communication. The public needs to understand the complexities of genetic engineering, its potential benefits and risks, in order to engage in informed discussions and shape its trajectory. This requires a concerted effort from scientists, policymakers, and the media to present information clearly, honestly, and accessibly. Ignoring public concerns or resorting to technical jargon only serves to fuel

mistrust and hinder progress. We should actively engage with the public, providing educational resources and fostering dialogue that considers ethical perspectives alongside scientific ones.

Furthermore, robust ethical frameworks are essential to guide the development and application of these technologies. These frameworks should not merely react to emerging issues; rather, they should proactively anticipate potential problems and establish clear guidelines for research and development. This requires interdisciplinary collaboration involving ethicists, scientists, legal scholars, and policymakers to navigate the complex ethical landscape. The aim should be to create a system of checks and balances that protects both individual rights and the common good.

Equitable access to the benefits of genetic engineering is paramount. The technologies we develop should not widen the gap between the rich and the poor, or exacerbate existing health disparities. Affordable and accessible healthcare solutions derived from genetic engineering must be prioritized, ensuring that these life-altering advancements are available to all, regardless of socioeconomic status or geographic location. We must carefully consider the potential for genetic technologies to be used for purposes of social engineering and eugenics. Such possibilities need to be proactively addressed and forestalled.

Investing in rigorous safety testing and risk assessment protocols is crucial for responsible innovation. The potential unintended consequences of genetic engineering, whether ecological or health-related, must be carefully considered and mitigated. Independent oversight bodies should be established to review research proposals, monitor ongoing studies, and ensure that appropriate safety measures are in place. This independent evaluation shouldn't simply be a formality, but an integral part of the innovation process.

Education plays a pivotal role in fostering responsible innovation. We must equip future generations with the critical thinking skills and scientific literacy needed to understand and engage with these complex issues. Integrating ethical considerations into science education at all levels is essential to cultivate a generation of scientists, policymakers, and citizens who are committed to the responsible development and application of genetic technologies. This includes fostering critical thinking about the societal implications of genetic engineering and understanding how these issues might impact us all differently.

The development of clear guidelines regarding intellectual property rights is also vital. Balancing the need to incentivize innovation with the need for broad access to life-saving technologies requires careful consideration. A system that promotes both innovation and equity is needed, and this involves a nuanced balancing of financial incentives and equitable distribution. This isn't merely about sharing profits—it's about ensuring that everyone benefits from these revolutionary advances in medicine.

Finally, international cooperation is essential. Genetic engineering has global implications, and its responsible development requires a concerted effort from the international community. Sharing best practices, harmonizing regulatory frameworks, and collaborating on research are all crucial steps in ensuring that genetic engineering benefits humanity as a whole. Failing to engage in meaningful international collaborations leaves a potential for abuse and uneven access. A cooperative and globally cohesive strategy is critical for success. The pursuit of responsible innovation in genetic engineering is not just a scientific endeavor; it's a moral imperative. It demands our constant vigilance, our collective wisdom, and our unwavering commitment to a future where these powerful technologies serve all of humanity.

www.ingramcontent.com/pod-product-compliance
Lightning Source LLC
LaVergne TN
LVHW022021200926
848481LV00007B/2476
* 9 7 9 8 3 4 9 2 8 5 8 0 6 *